What's So Great About Granite?

Jennifer H. Carey
Photographs by Marli Bryant Miller

2009
Mountain Press Publishing Company
Missoula, Montana

What's So
COOL About
GEOLOGY

The WHAT'S SO COOL ABOUT GEOLOGY series uses a lively and inviting tone and colorful photos and figures to introduce readers of all ages and levels of knowledge to geologic features you can see in the natural world.

Library of Congress Cataloging-in-Publication Data
Carey, Jennifer H., 1965–
  What's so great about granite? / Jennifer H. Carey and Marli Bryant Miller.
      p. cm.
  Includes bibliographical references and index.
  ISBN 978-0-87842-563-1 (pbk. : alk. paper)
  1. Granite—Popular works. I. Miller, Marli Bryant, 1960– II. Title.
  QE462.G7C37 2009
  552'.3—dc22
                                                                                  2009028442

PRINTED IN HONG KONG BY MANTEC PRODUCTION COMPANY

Mountain Press Publishing Company
P.O. Box 2399 • Missoula, Montana 59806
(406) 728-1900

Granite dome in Colorado.

Acknowledgments

I'd like to thank everyone at Mountain Press Publishing Company for their support, including history editor Gwen McKenna, whose flip comment led to the title of the book. Mountain Press editors James Lainsbury, Heath Weaver, and Beth Parker provided critical input on early drafts, as did geologist and author Wendell Duffield. James also edited the manuscript, and I included a photograph he took of strange circular features he found in granite while on a backpacking trip in the Bitterroot Mountains. Jasmine Star, a freelance editor, offered valuable suggestions on content in addition to her stellar copyediting. Numerous individuals across the country provided information about the places listed in the state-by-state appendix. My friend Becky MacDonald provided ideas and coffee during early brainstorming sessions, and artist Dorothy Norton captured my silly thoughts in cartoon form.

Though the book delves into complex topics, I tried to keep the technical jargon to a minimum and the tone light. I want readers of all ages and backgrounds to enjoy this celebration of granite.

*Granite and gneiss (dark stripe and pinnacles in background)
at Rocky Mountain National Park in Colorado.*

Contents

Granite domes at City of Rocks National Reserve in Idaho. —NATIONAL PARK SERVICE PHOTO

What's So Great About Granite?

What's so great about granite? You can walk on it barefoot, for starters, but it has many other wonderful qualities too. Unlike most other rocks, the average person can learn to recognize it from far away or up close. Many rocks are difficult to identify, even for geologists. They must get out their hand lens, bottle of acid, and hammer, and even take a sample back to the laboratory to analyze. But you can identify a granite outcrop from a car window at 60 miles per hour because of the distinctive shapes it forms as it erodes. It is also common, so everyone has a chance to see it. More than thirty states have outcrops of granite, many of which are dramatic features of national and state parks.

With endless variations in colors and patterns, granite is also beautiful. But its brilliance goes far beyond its sparkle. Granite is a major part of the foundation of our continent, the solid mass known as *bedrock* that lies beneath the soil and unconsolidated sediments deposited by wind and water.

Granite tells a geologic story about magmas and the hidden processes at work inside Earth. By studying granite, geologists can discover how continents form and what is happening deep beneath today's volcanoes. Although some geologists have spent a lifetime studying the intricate details of this rock, there is still much to learn.

You can play on granite, climb on it, and sleep on it. Yes! Even sleep on it. Its smooth surface gives new meaning to the term *bedrock*.

Water collects in pockets on granite bedrock in Colorado.

Mount Whitney, the tallest mountain in the contiguous United States, is granite.

What is Granite?

If I asked if you've ever seen graywacke, andesite, or phyllite, you might not know I was talking about rocks. But you've probably heard of granite. It may even be the first rock you learned by name, perhaps because it forms towering cliffs and peaks in well-known places, such as Yosemite National Park, Grand Teton National Park, and Mount Rushmore National Memorial. Or maybe it's because granite is a beautiful rock used for floors, countertops, building stones, statues, and gravestones. Either way, it's a common rock easily taken for granted.

Granite was used as facing on this office building.

Castle Crags State Park in northern California features pinnacles of granite.

Many of the last wild places in North America feature granite because the rough topography formed by this hard rock discouraged settlers. You can't plow granite. You can't dig holes in granite for fence posts, wells, root cellars, or outhouses.

Even if you live on the Great Plains or the coastal plain of the southeastern United States, you're not far from granite. It shapes many landscapes in North America, providing the backbone of ridges and mountain peaks and the hard walls of river canyons. Rivers carry pebbles and cobbles downstream, so you may see granite in a river bottom or on a pebble beach many miles from its source. A glacier may even have carried a giant granite boulder and dropped it smack-dab in the middle of a prairie long ago, when ice covered much of the northern states and brought rocks south from Canada.

A granite boulder rests in a grassy field in Yellowstone National Park. A glacier moved it many miles from its original location.

But what exactly is granite, how does it form, and what can it tell us about the inner workings of Earth?

Sparkles and Speckles: The Minerals of Granite

If you find a whitish, grayish, or pinkish rock that is speckled and reflects sunlight, then you've probably found granite. Granite is made up of varying amounts of five main minerals. The minerals quartz, potassium feldspar, and plagioclase feldspar are light colored and make up the bulk of the rock.

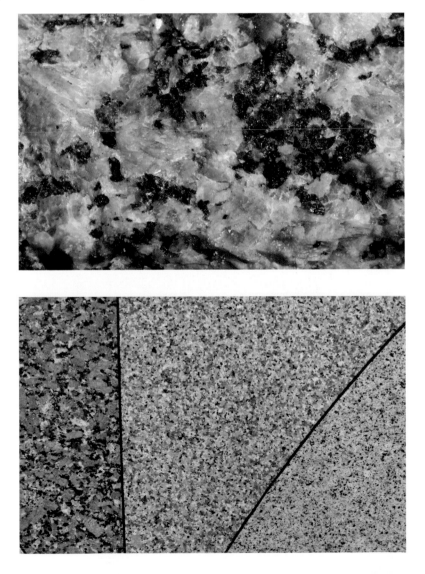

Granite is a speckled rock that reflects light.

This decorative floor features different varieties of granite.

Smaller amounts of the minerals mica (usually biotite but sometimes muscovite) and hornblende (sometimes called *amphibole*) provide the dark speckles. Mineral crystals in many rocks are too small to see without a microscope, but those in granite are usually visible to the naked eye.

The flat crystal faces of the minerals in granite give it that dazzling sparkle. Minerals, the building blocks of rocks, have crystal structures made of atoms arranged in geometric patterns. This orderly arrangement creates the straight lines and flat surfaces that make crystals so beautiful, and sometimes so valuable (think diamonds). One of the simplest crystal structures is a cube; both common table salt and pyrite, also known as *fool's gold*, are cubes. The six sides of the cube are its crystal faces. Most mineral crystals in granite are more complex than a cube. For instance, quartz is a six-sided prism with pointed ends.

Granite forms deep beneath Earth's surface as minerals crystallize slowly from magma, a hot fluid that is melted rock. As the magma cools, the different crystals form. In granite, the mineral crystals crowd into each other as they grow, forming an interlocking arrangement of imperfectly shaped crystals. Perfect crystal shapes don't usually get a chance to form in granite because of the crowding.

When you look through a microscope at an extremely thin slice of granite, you can see the interlocking nature of its minerals.

When you look at the fresh surface of a recently broken chunk of granite, you see broken minerals. Some minerals break along zones of weakness within their crystal structure, often parallel to the crystal faces, so the break resembles the mineral's particular crystal structure. Mica breaks along one zone of weakness and looks like a page in a tiny book, or a flake. Hornblende breaks to look like a needle. Feldspar has two dominant breaks approximately at right angles to each other, so its crystals look boxy. On the

Feldspar generally breaks along two dominant surfaces, so it looks boxy.

When quartz has room to grow in holes or cracks in a rock, it may form diamond-like crystals. In granite, on the other hand, it may not have perfect crystal shapes because it formed in cooling magma, where it competed for space with other minerals that were growing.

The micas biotite (left) and muscovite (right) break along one flat surface.

Look for flat needlelike surfaces to identify the dark mineral hornblende.

When the mineral quartz breaks, it leaves a rounded surface called a conchoidal fracture.

other hand, the crystalline structure of quartz is equally strong in all directions. It has no zones of weakness, so it doesn't break along flat surfaces. Breaks in quartz leave a smooth, curved surface called a *conchoidal fracture*. Glass also breaks this way. When you see a mineral grain in granite that looks like a piece of broken glass, it's quartz. And it can be sharp.

Rock climbers have a love-hate relationship with the sharp edges of quartz grains. Those edges provide the friction necessary for the rubber sole of a climbing shoe and the skin of a climber's fingertips to cling to a granite wall—a good thing if you want to hold on, but a bad thing if you fall and scrape against the rock.

The exact colors in a chunk of granite depend on the colors of the minerals in it. The light-colored minerals can be gray, white, or pink. The dark minerals can be dark green to black to brown. A mineral's color varies according to its chemical composition, or what was in the melted rock to begin with, but more on that later. In general, granite is a light-colored rock with dark speckles and interlocking crystals you can see without a microscope.

Different varieties of quartz have different colors. The purple variety is called amethyst. Other colors are informally called "smoky" and "rose."

ANIMAL, MINERAL, OR VEGETABLE?

In the guessing game Animal, Mineral, or Vegetable?, one player thinks of an object and the other players try to guess what it is by asking yes or no questions. The guessers usually begin with "Is it an animal?" "Is it a mineral?" or "Is it a vegetable?" In the game, a mineral could be a rock or a mountain or an element, such as gold. It's anything that isn't alive or made of something that once was alive.

In the world of earth science, the word *mineral* has a more specific meaning. Minerals are the building blocks of rocks. A mineral has a characteristic chemical composition made of certain elements, or atoms, in specific proportions. Atoms are the particles that make up all things in the universe, including us. The atoms in a mineral are arranged in repeating geometric patterns that give each mineral its characteristic crystal shape, color, and hardness. A mineral's hardness, or its resistance to being scratched, is directly related to the strength of the bonds between the atoms in the mineral. Hardness is rated on a scale from 1 to 10, with 1 being the softest, such as talc, and 10 the hardest, such as diamond. Quartz has a hardness of 7.

Pigeonholing Rocks: This Is Granite and So Is That

Rocks cannot be classified as precisely as living things. If you and a friend see a large dark object with a white head dive out of a tree and snatch a fish from the water, you'll quickly agree it's an animal, most certainly a bird. You may even identify what species it is, perhaps a bald eagle. Imagine, though, if you were still trying to decide if it was a plant or an animal? That's more like the rock identification process. Sure, there are living things that don't fit neatly into categories, especially at the species level, but most can be distinguished as a plant or an animal with a single glance.

Geologists classify rocks in one of three categories: *igneous, sedimentary*, and *metamorphic*. Igneous rocks solidify from melted rock. Sedimentary rocks are layers of sediment, such as sand and mud, that were deposited by water, wind, or ice and then hardened. Metamorphic rocks are igneous or sedimentary rocks that were changed by heat and pressure. So in what category would you place a rock composed of volcanic ash that settled out of the air into a thin layer on the ground? Because it started as magma but was deposited by wind, that's a tricky situation, and perhaps we should simply call it a volcanic deposit!

Some rocks are mixtures of several rock types, further complicating the identification process. Sandstone, composed of sand-sized grains of sediment, and siltstone, composed of silt-sized grains, are two distinct rock types that were originally sediments deposited by water or wind. However, a sandstone rock may contain a silt-rich layer, and a siltstone rock may contain a sand-rich layer. They're classified based on the size of the majority of the grains. Granites can also vary, depending on their mineral constituents. A piece of granite on one side of a boulder may not look exactly like a piece on the other side of the same boulder. Despite these complications, geologists have sorted through the confusion and provided some guidance for rock identification beyond the three main categories.

Once you've determined whether a rock is igneous, sedimentary, or metamorphic, the next thing to look at is grain size or crystal size. To classify a sedimentary rock, you must determine whether it's made of tiny grains like clay (making it shale), small grains like silt (making it siltstone), medium grains like sand (making it sandstone), large grains like cobbles (making it conglomerate), or some combination of sizes. The metamorphic rocks slate, phyllite, and schist are also organized by grain size, with slate having the smallest grains and schist the largest.

For igneous rocks, you must determine whether you can see the mineral crystals. If you can't, it's an extrusive igneous rock, which means it solidified on the surface after magma exploded from a volcano or oozed from a crack in the ground. In this case, the magma cooled too quickly for crystals to grow big enough to see, and sometimes crystals don't grow at all. Some extrusive igneous rocks do contain mineral crystals large enough to see, but they're usually a minor component of the rock. These crystals formed beneath the surface before the magma was erupted. However, if the surface seems to be composed mostly of visible crystals, it's an intrusive igneous rock, meaning it solidified slowly below the surface in an environment in which crystals could grow big enough to see. Granite is an intrusive igneous rock.

Intrusive rocks are further divided into three main types—granite, diorite, and gabbro—distinguished by the amount of silica they contain. Silica is a molecule composed of one atom of the element silicon and two atoms of oxygen. Most minerals that crystallize from magma are called *silicates*

IGNEOUS ROCKS

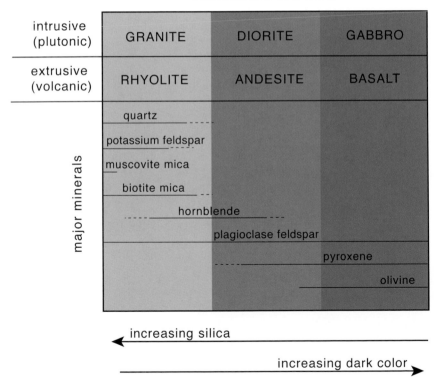

Igneous rocks are first divided into two broad groups, extrusive and intrusive. These two groups are further subdivided into three main rock types based on silica content. The classification scheme for igneous rocks presented here is a very simplified overview.

because they contain a compound—or union—of the elements silicon and oxygen. Granite is approximately 70 percent silica.

How can the average person tell how much silica a rock has? It just so happens that the more silica an intrusive igneous rock has, the lighter its color. Silica is a major component of feldspar, and the mineral quartz is 100 percent silica. Both are usually light colored. Granite has at least 10 percent quartz and usually more than 20 percent. The dark minerals, mica and hornblende, are also silicates, but they don't contain as much silica as the light minerals.

When working in the field, geologists determine if an intrusive igneous rock is granite, diorite, or gabbro by estimating the percentage of dark minerals in the rock. Mineral color can be deceiving, so this is just a rough estimation. Granite typically has less than 15 percent dark minerals, diorite between 15 and 40 percent, and gabbro between 40 and 80 percent. Diorite

Granite (left), *diorite* (middle), *and gabbro* (right).

TRUE GRANITE

Many professional geologists would look at this book and say, "This book isn't about granite, it's about *granitoids*." It's true that within detailed schemes that attempt to classify all quartz- and feldspar-rich intrusive igneous rocks, the term *granite* is reserved for a specific rock type in which the percentage of each mineral it contains falls within a specified range. True granite is just one of a group of granitic rocks called *granitoids*. One of the more common granitoids is granodiorite, a major component of the Sierra Nevada. Unlike true granite, it has a lot more plagioclase feldspar than potassium feldspar. I call granitoids "granite" in this book because the distinction is unimportant to the casual observer. I also include cousins of true granite such as quartz monzonite, the granitic

rock that Stone Mountain in Georgia is composed of. It has less quartz than true granite, but it's still an intrusive igneous rock dominated by feldspar and quartz.

The classification of granitoids continues to evolve, and the more recently developed classification schemes rely more on the element content of the entire rock than the mineral proportions. Geologists grind up a rock sample, which destroys the minerals, and then analyze the powder with a sophisticated machine. They can determine the exact percentages of common elements (iron, calcium, sodium, magnesium, silicon, aluminum, potassium, phosphorus, titanium, and manganese) and more unusual elements, known as *trace elements*, the sample contains.

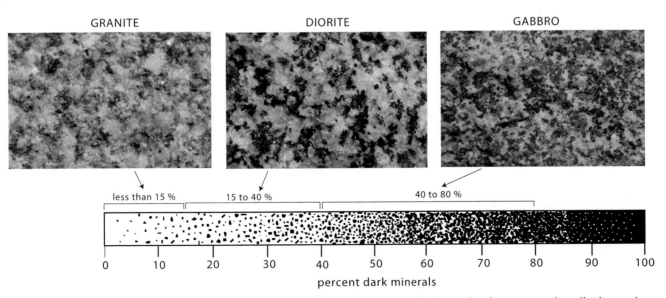

GRANITE   DIORITE   GABBRO

less than 15 %   15 to 40 %   40 to 80 %

0   10   20   30   40   50   60   70   80   90   100

percent dark minerals

Granite has fewer dark minerals than the other two main intrusive igneous rocks, diorite and gabbro. The bar gives an idea of the percentage of dark minerals. The lighter-colored minerals contain more silica than the darker minerals.

has very little if any quartz, and gabbro has none. Quarried stone called "black granite" is usually gabbro. In addition to hornblende, diorite and gabbro contain pyroxene and olivine, dark minerals that are never present in granite.

The Pretenders: What Is Not Granite

A handful of rocks look enough like granite to be confused with it, but some of them aren't even igneous.

Sandstone can have a resemblance to granite, but it's only skin-deep. Sandstone is a sedimentary rock. Calling it granite is like calling rabbitbrush an animal rather than a plant. Some sandstone looks similar to granite because it's often made of grains of quartz and feldspar, the same minerals that occur in granite. In fact, the minerals may have originally been part of a granite rock but were broken into grains and then rolled and tossed by water or wind until they became round grains of sand. The mineral grains in sandstone don't interlock like those in granite; they're held together with natural cement, a mineral material that precipitated in water, such as calcium carbonate or iron oxide. **Quartzite**, sandstone changed by heat and pressure, can also look like granite, but it is a metamorphic rock.

This sandstone contains rounded grains of quartz.

Rhyolite, an extrusive igneous rock, has the same chemical composition as granite. However, rhyolite cooled so quickly on the surface after erupting from a volcano that the majority of its crystals didn't have much time to grow. Most are so small that they can't be seen with the naked eye. Sometimes bigger crystals had already formed in the magma before it reached the surface. In those cases you'll be able to see them, but they're surrounded by crystals too small to see, as well as by glass—magma that cooled so fast that no crystalline structure formed at all. Had the magma cooled deep below the surface, it would be granite.

This rhyolite has rectangular holes that contained mineral crystals that weathered away.

WHEN GRANITE IS GNEISS

Gneiss (pronounced *nice*) is a metamorphic rock with alternating light and dark layers and visible mineral grains. The light layers may look like granite. Gneisses form when sedimentary rocks or igneous rocks, including granite, are subjected to intense heat and pressure. A metamorphosed granite is often called *granite gneiss*. It has the same mineral composition as granite but also has gneiss's characteristic bands, which can be well-defined layers or just a faint alignment of the crystals of flat biotite and needlelike hornblende.

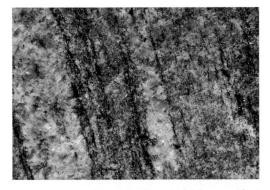

This gneiss has light layers that look like granite.

Where Can You Find Granite?

Some of the highest mountain ranges in the world, including the Himalayas, Andes, and Rockies, contain a lot of granite. Granite is often found in mountains for two main reasons: First, granite is a major product of the tectonic plate collisions that create mountains. Second, granite is a hard rock that resists erosional forces such as wind, water, and ice. Its resistance is a function of its very uniform nature and its interlocking crystalline minerals, including the hard mineral quartz. Layers, which are common in sedimentary and metamorphosed sedimentary rocks, create avenues for water and ice to exert erosional effects, but granite has no layers, hence its uniformity. The softer rock surrounding granite erodes away but the granite remains, protruding above its surroundings. As a river erodes a channel, it usually finds a path around a body of granite rather than cutting through it. When possible, engineers will even route a road around a granitic body rather than drill and blast the road through it.

*Mount Whitney, the highest mountain in the contiguous United States, is entirely granite.
The Alabama Hills in the foreground are also granite.*

Low-lying granite in northern Wisconsin was part of a mountain range more than 1 billion years ago. The granite has been substantially eroded in that time.

Not all granite masses are tall mountains today. Given enough time, even granite erodes. Flat to hilly regions of granite are places where mountains existed hundreds of millions or even billions of years ago. The Boundary Waters Canoe Area Wilderness in northern Minnesota is one such place.

Coast Range

CANADA

The region of granitic bedrock in Canada and the upper Midwest includes a considerable amount of gneiss.

Bitterroot Range

Black Hills

Rockies

Sierra Nevada

UNITED STATES

Appalachians

Peninsular Ranges

MEXICO

N

200 miles

Granite bodies in North America coincide with modern and ancient mountain ranges.

Hard-Boiled Tectonics

Earth's surface resembles the cracked shell of a hard-boiled egg. This shell, or *crust* as it is known, is made of large, relatively thin pieces of rock called *tectonic plates*. These rigid plates ride around on the mantle, the egg white of our egg model. The mantle surrounds Earth's core, which corresponds to the egg yolk.

The core is extremely hot—more than 8,500 degrees Fahrenheit. The outer part of the core is magma. The heat from the core drives the movement of the tectonic plates. Since heat flows toward cooler areas, heat from the core moves outward into the mantle, just as the heat from a stove top flows into a pot of water. Flowing currents, called *convection cells*, develop in a pot of near-boiling water, driven upward by the heat of the stove and

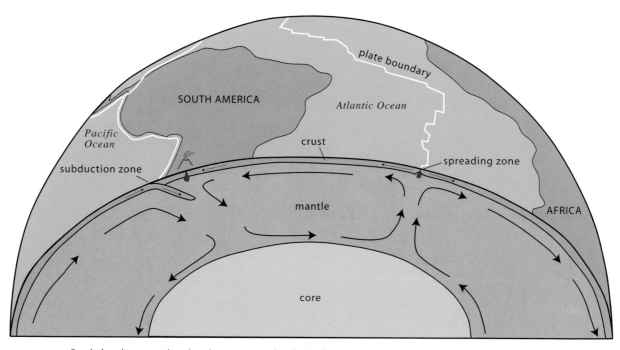

Rock in the mantle circulates very slowly in large convection cells, transporting heat toward the surface and driving the motion of the crust's tectonic plates.

the relative coolness of the overlying water. The currents rise to the surface of the pot, generating bubbles and steam, and cooler surface water sinks to the bottom of the pot. Similar currents transport heat inside Earth. The warmed mantle is solid rock but flows slowly, like warm taffy. When the upward-moving mantle rock reaches the upper mantle, it cools. This cooled mantle rock sinks back toward the core, perpetuating the cycle of currents. These currents are what move the plates along the surface of the planet.

Earth's crust, the eggshell, ranges from about 4 miles thick to about 40 miles thick. Thinner oceanic crust forms where magma from the mantle rises to the surface in the middle of oceans at underwater spreading zones, places where tectonic plates are moving away from each other. The magma that erupts from the spreading zones quickly solidifies into a dark, dense rock called *basalt*.

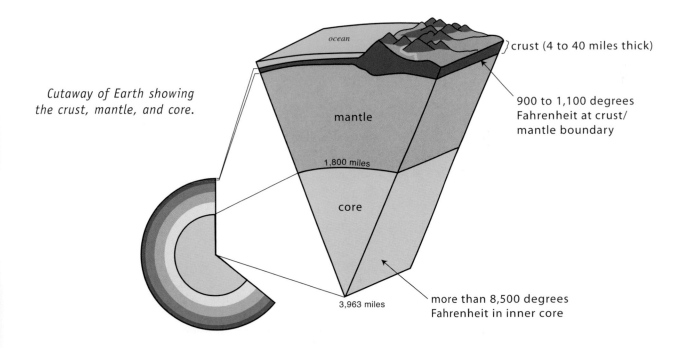

Cutaway of Earth showing the crust, mantle, and core.

ocean

crust (4 to 40 miles thick)

900 to 1,100 degrees Fahrenheit at crust/ mantle boundary

mantle

1,800 miles

core

3,963 miles

more than 8,500 degrees Fahrenheit in inner core

The thicker continental crust is a jumble of all types of rock, including large amounts of granite. Continental crust is less dense than oceanic crust, so where plates of the two types meet, the opposite setting of a spreading zone, continental crust rises up and over the oceanic crust. This type of tectonic margin is called a *subduction zone*. At the surface, rocks are crumpled into mountains and pieces of the ocean floor get plastered onto and shoved under the continent. To visualize the continental margin at a subduction zone, imagine what happens to vegetables, jars, and boxes on the small conveyor belt in the checkout lane at a grocery store. When they reach the edge of the counter, where the belt dives below it, the groceries bump into each other, turn, and pile up. Continental crust is thickest where mountains exist, such as above a subduction zone.

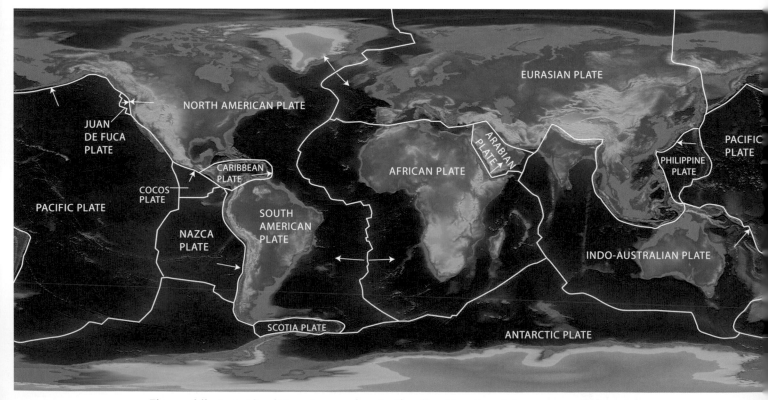

The world's tectonic plates. Arrows denote the direction the plates are currently moving.

ARE YOU DENSE?

Density is a measure of how much a substance weighs per a specific volume. Let's compare a solid piece of granite with pumice, a cavity-filled rock with a structure that resembles a sponge. Pumice solidifies from lava blown out of the top of a volcano. Of course, a pound of granite weighs the same as a pound of pumice. But that pound of granite will sink in water, whereas the pumice can actually float, at least until water manages to seep into all of its holes. This is because the pound of pumice, being full of holes, occupies considerably more space. Basically, the pound of granite would be about the size of a baseball, while the pound of pumice would be about the size of a softball.

DOROTHY NORTON

"This book is dense."

Volcanoes form above subduction zones. Beneath them lie areas of magma, some of which cool into granite deep within the crust. Whether the granite will ever be exposed at the surface depends on geologic events that occur over millions of years. Linear bands of granite at Earth's surface, such as the Sierra Nevada in California, mark areas where subduction zones existed long ago.

Where two continents collide, neither gives way, so huge mountains form. The highest mountains in the world, the Himalayas, began to form 50 million years ago when India collided with Asia—a slow-motion collision that continues to this day.

Upwardly Mobile

Many high mountains, including 14,259-foot Longs Peak in Colorado and 13,770-foot Grand Teton in Wyoming, are granite. If granite solidifies miles below the surface, you may wonder how it ends up at the top of mountains. A giant chunk of granite can't exactly crawl through the rock surrounding it. The theory of plate tectonics provides the answer.

Where tectonic plates collide, the crust bunches up and thickens, like the front end of a car that has crumpled up in an accident. Because the mantle flows like taffy, it's pushed down by the extra weight of the bunched-up crust. The top of the crust, however, is still higher than it was before these tectonic events, and it is this crust that forms mountains at Earth's surface.

Over time, rain, wind, glaciers, and other forces erode the mountains, carrying sediment into valleys and out to sea. As more and more material is removed by erosion, the crust gets progressively thinner and less heavy, so the mantle below it begins to rise back up. The portion of the crust that had been pushed down into the mantle also rises, sometimes thousands of feet. This newly exposed part of the crust may include granite, which now lies much higher than where it formed. Mountains of granite also rise along faults, large breaks in the crust. Movement between and within tectonic plates causes the crust to move along faults.

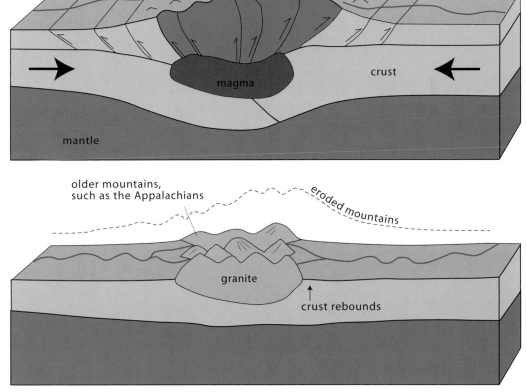

Granite that forms deep within the crust may someday be exposed at the surface.

huge mountains, such as the Himalayas

Where plates collide, huge mountains form and underlying rock melts, becoming magma in the lower crust. Arrows denote the direction of plate movement. Over time the magma crystallizes as granite.

crust

magma

mantle

older mountains, such as the Appalachians

eroded mountains

granite

crust rebounds

As mountains erode, the crust rebounds and the granite is exposed in mountains.

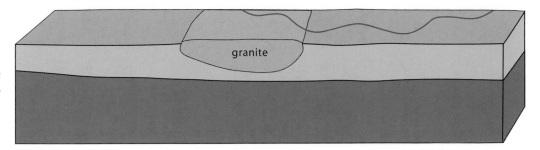

granite

Flat-lying or hilly regions of granite are places where mountains once existed.

28

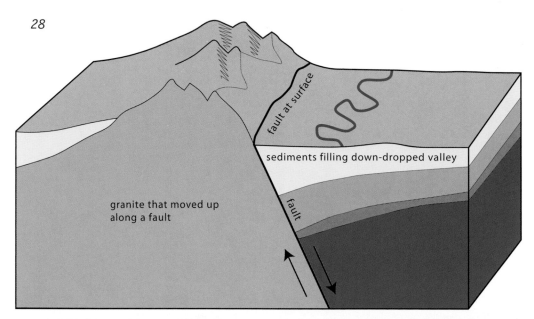

granite that moved up
along a fault

fault at surface

sediments filling down-dropped valley

fault

Some mountains rise along faults. Arrows denote the direction of movement.

The granite and gneiss of the Teton Range in Wyoming have risen thousands of feet along a fault while the valley floor of Jackson Hole has dropped. Overall, the total vertical displacement along the fault may be 33,000 feet.

FLOATERS IN THE BATHTUB

The tectonic process involved in raising granite miles and miles to Earth's surface can be difficult to comprehend. You can create a model with two short pieces of two-by-six lumber and a bathtub full of water. If you put one piece of lumber in the tub, the wood will sink until it displaces a volume of water equal to its own weight. But because wood is usually less dense than water, it won't sink completely. This piece of wood represents the mountains that form as crust piles up at a collision zone, as well as the sinking of the crust and depression of the mantle (the water). As with the wood on the water, the crust is less dense than the mantle, so it only sinks so far.

If you put the second piece of two-by-six on top of the first piece to represent further piling up of crust and more mountain building, the first piece will sink lower, displacing more water. But the top of the second piece of wood will be higher than the first piece was alone. The two pieces and the water, like the pieces of crust that accumulate on top of the mantle due to plate collision, reach a new equilibrium, or balance. If you then remove the top piece, the first piece will rise higher, just as the crust rises as mountains erode and the mantle beneath it rebounds.

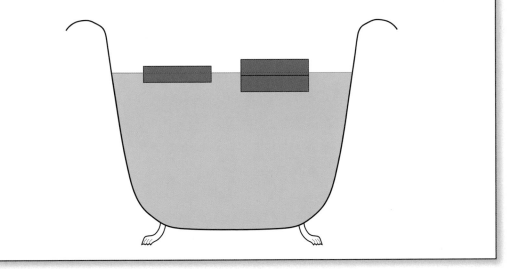

How Does Granite Form?

How granite forms has been hotly debated for centuries. In the 1700s, some geologists argued that granite precipitated out of ocean water. In the early to mid 1900s, some argued that crustal rock changed into granite without completely melting—transforming due to the movement of elements within a rock heated during metamorphism. All along there were also scientists who thought granite solidified from magma below the surface, and they were right. But what exactly is going on down there? Part of Earth's core is magma, but the rest of Earth is solid. Where does the magma that forms granite come from? And what, precisely, is this substance?

Magma is melted rock, a liquid that may contain gases and mineral crystals suspended in it. Where it oozes onto the surface, this red-hot, flowing liquid is called lava, but when it's inside Earth it's called magma. All extrusive igneous rocks solidify from lava, and all intrusive igneous rocks solidify from magma. It's best not to lose sleep over this odd distinction.

Lava flow in Hawaii.

Magma inside Earth is under intense pressure. In addition to the pressure created by the mass of overlying rock and movement of tectonic plates, magma occupies more space than the solid rock from which it melted. Because it's hot, magma's atoms aren't held as tightly together as those in the minerals of cooler, denser, solid rock. In essence, magma is being squeezed on all sides. If possible, it will escape through fractures in Earth's crust to a place where there's less pressure. Imagine if you squeezed a lemon really hard in your hand or with a vise. Eventually the juice would come squirting out, like lava from a volcano.

So why isn't lava shooting out of volcanoes in fiery plumes everywhere you look? Because for the most part, even with the high temperatures that occur below the surface, the crust and the mantle are solid. (Even though it flows like taffy, the mantle is still a solid.) The high pressure from the overlying weight of miles and miles of rock keeps solid mineral crystals from melting and becoming liquid. Pressure is one of the important factors determining the temperature at which water, rock, or any other substance changes from solid to liquid and from liquid to gas. For example, the boiling point of a substance, such as water, is the temperature at which it changes from a liquid to a gas. Many of us know that water boils at a higher temperature at sea level than in the mountains but do not know why. The answer is that there is more atmospheric pressure at sea level; therefore, more heat, or energy, is needed to change the water molecules into gas.

Pressure also affects the temperature at which rocks melt. At Earth's surface, granite melts at temperatures between 1,200 and 1,600 degrees Fahrenheit, but it would still be solid at more than 1,800 degrees Fahrenheit if it were buried under 50 miles of crust. More heat, or energy, is needed to change its solid minerals into a liquid. So how does solid crust or mantle melt and form magma? One of three things must happen: there must be a decrease in pressure, an increase in heat, or the addition of water. Different things can happen within Earth to bring these changes about. Since they are such abstract concepts, it's best to use the following figure to explain them.

Several melting scenarios. Arrows denote the direction of movement.

1. *Where two tectonic plates collide, the crust thickens and depresses the mantle. Some of the rock in the lower part of the crust may melt because it's pushed into a zone of higher temperatures. (Temperature increases with depth in Earth.)*

2. *The warm mantle, heated by Earth's core, rises toward the surface. The pressure decreases as it rises because there is less rock above it, so some of the rising mantle melts.*

3. *Magma may stop rising at a natural barrier. There is a boundary near the top of the mantle that separates rock that flows like taffy from overlying, more brittle rock. Geologists believe that magma can be trapped beneath the boundary. Magma may also become trapped beneath rock that doesn't have fractures for the magma to flow through. In either situation, heat from the pooled magma can melt rock above it, creating more magma.*

4. *Water circulating through the crust can break the chemical bonds that hold atoms together in mineral crystals, essentially lowering a rock's melting temperature. Clay and other minerals that accumulated on the floor of the ocean for millions of years contain water in their molecular structure. Where a plate of oceanic crust slips below another plate, these minerals become unstable as they move deeper into the Earth with the down-going plate. Water is released from the minerals and migrates upward, lowering the melting temperature of the rock it moves through.*

Once magma has formed, it can migrate. Or it can stay put. No matter what, eventually it will cool and become solid rock again.

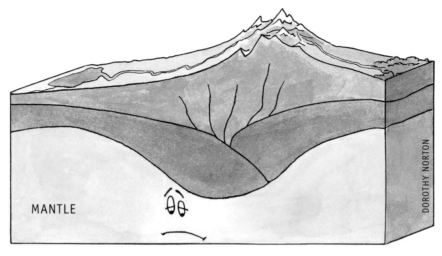

"I'm so depressed."

Am I Intruding?

If you walk into a room where several people are having a quiet conversation, you might be intruding. When magma moves into part of Earth's crust it is said to be intruding, and the body of rock that solidifies from the magma is called an *intrusion*. Magma intrusions can take 1,000 to 100,000 years to solidify and typically develop 1.5 to 30 miles beneath the surface. It's easy to understand why they're called *intrusions* when you see some of the curious relationships between them and the rock they intruded, which is called *country rock*. In some places long tentacles of intrusive rock reach into country rock. In other places, the intrusion pushed through fractures, deforming the country rock and prying off pieces of it. All bodies of granite are intrusions.

This contact between the country rock (dark gray) and granite (pinkish white) is complex.

Small fingers of granite intruded this country rock.

The light-colored granite pluton (below) intruded dark volcanic rock (above).

Intrusions form when magma stops moving and solidifies. The magma may stop when it reaches cooler regions of the crust, where it loses heat to the surrounding rock or groundwater. Or the magma may become trapped by an impenetrable layer, such as a rock with few fractures. Since the magma has nowhere to go, it pools below the barrier and solidifies.

One perplexing question related to granite has baffled geologists and nongeologists alike for a long time. How could the magma of a large intrusion, such as a pluton, fit anywhere beneath Earth's surface without displacing something? There aren't large holes down there that could fill with magma. Geologists know that the crust bulges up in areas with active volcanoes, such as Yellowstone National Park, but probably not enough to accommodate a large pluton. It's possible that in a volcanic region, granite magma could rise and replace the lava that was spewed out the top of a volcano, much like water moves into a well to replace the water that was pumped out. But it's hard to imagine these mechanisms creating the space needed for a large pluton.

As magma flowed into a crack, it plucked off pieces of country rock before solidifying as this granite dike.

BODIES OF THE UNDERWORLD

A body of granite can be as narrow as your pinkie finger or more than 100 miles wide. A general term for any igneous body of rock is *pluton*, named for Pluto, the Greek god of the underworld. A pluton cools and solidifies from a single pool of magma. A large region of igneous rock, such as that which composes the bulk of the Sierra Nevada in California or the Bitterroot Range in Idaho, is called a *batholith*. Batholiths can be more than 100 miles long and 50 miles wide and usually consist of many individual plutons of different ages and compositions. Igneous bodies of different shapes and sizes have specific names. For example, a narrow body that forms when magma fills a fracture in rock is called a *dike*. If the narrow body forms between layers of rock, it's called a *sill*.

This granite sill cuts through gneiss.

A narrow granitic dike made entirely of light-colored minerals cuts through granite.

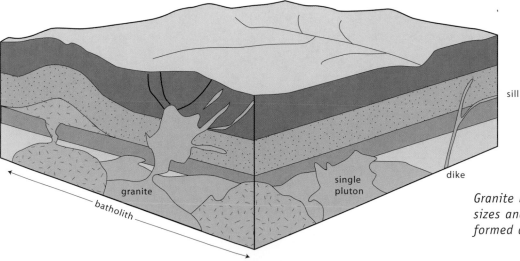

sill

dike

granite

single pluton

batholith

Granite bodies come in many sizes and shapes; all of them formed deep in Earth's crust.

The Sierra Nevada in California is a large batholith.

Movement of the solidifying granite magma broke this dark xenolith.

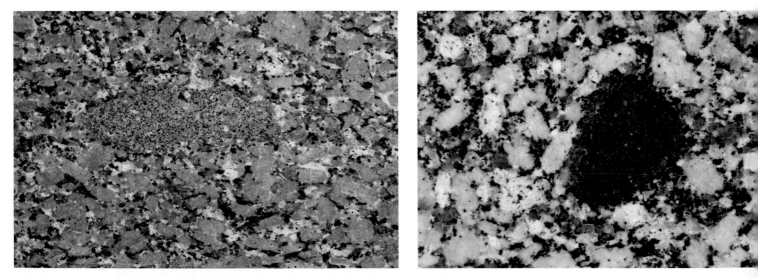

These dark xenoliths were partially absorbed by the magma that solidified as the surrounding granite.

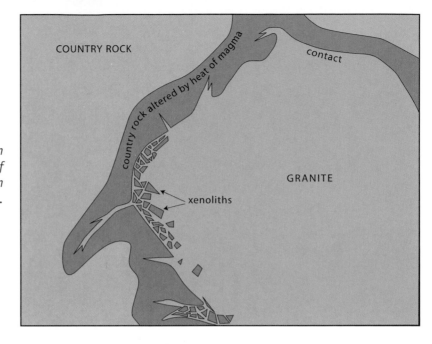

COUNTRY ROCK

country rock altered by heat of magma

contact

GRANITE

xenoliths

Country rock surrounds an intrusion of granite. Pieces of country rock enclosed within granite are called xenoliths.

COUNTRY ROCK: IT'S NOT MUSIC

Despite what the dictionary says, to a geologist *country rock* is not "rock music infused with country music." But it does involve infusing one thing with another. Country rock is the rock that magma intrudes. It's the rock surrounding a body of granite or that a granite dike cuts through.

When magma engulfs adjacent rock or moves forcefully toward the surface through cracks in the crust, it breaks off chunks of the surrounding country rock. Some of these chunks melt and become part of the hot mush, but others only melt along their edges before the magma solidifies around them. These pieces of rock, called *xenoliths*, are easy to see if they are distinctly different than the enclosing granite. Their Greek name suits them well; *xeno* means "stranger" and *lith* means "stone."

Although country rock isn't a genre of music, a xenolith is definitely music to a geologist's ears. It may be the only piece of evidence a geologist has to determine what rock the granite intruded and whether the granite is older or younger than another rock in the region. Understanding the relationship between country rock and intruding granite is important for geologists trying to establish the geologic history of a region, but the contact between the two is often hidden from sight, covered with younger sediment or volcanic rock.

One of the best treats of all time (second only to ice cream) is to find the place of contact between an intrusion of granite and country rock. There you can see how the heat of the magma changed the country rock. If the country rock was originally a soft sedimentary rock, it might be highly altered. Perhaps it was baked as hard as the adjacent granite and developed metamorphic minerals, such as garnet, that grew as the rock was heated. The farther away from the granite you get—in some places just a few feet—the more the country rock will resemble its old self.

Another mechanism that could help mitigate the space issue is if at least some of the magma of a pluton melted directly from the country rock. In this case, the additional magma would simply replace the space occupied by the rock it melted from. The existence of rocks known as *migmatites* in close proximity to granite is strong evidence in support of this scenario. Migmatites are a type of rock made up of a mixture of igneous rock (usually quartz- and feldspar-rich granite) and country rock. They may represent the margins of a former magma body, where the country rock didn't fully melt and become incorporated into the magma body.

A migmatite composed of light-colored granitic rock and country rock in the North Cascades, Washington.

DOROTHY NORTON

Country rock

Magma Soup

Not all magmas are the same. Imagine magma as a molten soup with many different ingredients. The soup's taste and texture depend on the type and quantity of each ingredient, how long each ingredient cooks, and how well the soup is mixed before serving. Likewise, the chemistry and texture of an intrusive igneous rock depend on the ingredients and "cooking" history of the magma soup it was derived from.

The most important factor in determining what kind of solidified igneous rock a magma will be is what the magma melted from—in other words, what are its ingredients? Some magmas melt from mantle rock, some from crust rock, and still others from a combination of the two. Specific minerals melt at specific temperatures and pressures, just as ice melts at 32 degrees Fahrenheit at the pressure at Earth's surface. Most rocks contain a combination of minerals, and as they begin to melt, the minerals with lower melting temperatures turn to magma first. For instance, in igneous rocks, quartz (100 percent silica), potassium feldspar, and sodium-rich plagioclase feldspar melt at lower temperatures than calcium-rich plagioclase feldspar and iron- and magnesium-rich minerals, such as hornblende, pyroxene, and olivine. So every magma that melted from an igneous rock is initially richer in silica, potassium, and sodium than the rock that is melting.

If a rock doesn't melt completely, the magma will have a different chemistry than the rock it melted from. When magma derived from partially melted mantle rock erupts at the surface, it solidifies as basalt, a rock with slightly more silica than the mantle. Magmas that melt from silica-rich continental crust will solidify as silica-rich rocks, such as granite. The source of the magma plays the biggest role in what the cooled rock will be, just as the ingredients determine the type of soup. But things can happen during the crystallization process that affect what the final rock will be.

After cooking the soup, the cook might remove something before serving it, perhaps skimming off a layer of fat from the surface or removing a bay leaf or soup bone. Certain ingredients may be removed from magma, too. As the magma cools, the heaviest mineral crystals begin forming first. These crystals, usually richer in iron and magnesium than the magma, can settle out of the magma soup to the bottom of the pool of magma. They may also be removed as the magma flows upward. Just as twigs get stuck in an eddy as the current of a river flows by, crystals can be left behind in the portion of magma that doesn't flow. With the early-forming crystals removed, the remaining magma solidifies as a type of rock that's different from what would have formed if the initial magma hadn't lost any of its minerals.

DOROTHY NORTON

"I think it needs a little more silica."

Reading Crystals

Individual crystals in granite often contain visible clues about what conditions were like deep within Earth as the magma solidified. Mineral growth is all about the pursuit of equilibrium, the process of reaching a balance with the surrounding environment. Growth depends on the availability of the elements a mineral needs, the availability of space, the temperature, and the pressure—all of which are usually changing.

Plagioclase feldspar, one of the main minerals in granite, can contain varying proportions of calcium and sodium. Calcium-rich plagioclase feldspar crystallizes at higher temperatures than sodium-rich plagioclase feldspar, so as magma cools, plagioclase feldspar crystals gradually change composition. Each new layer of growth may contain a bit more sodium and a bit less calcium. In fact, a single crystal may have a calcium-rich center and a sodium-rich exterior. If the temperature fluctuated as the crystals formed, the chemical composition of each layer would reflect the change in temperature. This change in the chemical makeup of an individual crystal is called *zoning*. Scientists can use zoned crystals to map out the cooling history of a particular magma. In many cases, you can even see the concentric rectangles of the zones with the naked eye.

Some granitic rocks have large individual crystals surrounded by a matrix of smaller crystals. The large crystals formed while the magma was cooling slowly. Since the individual crystals were surrounded by liquid magma, they had ample room to grow and thus developed characteristic crystal shapes. The matrix formed as the remaining magma cooled much faster, perhaps because it rose closer to the surface, where the temperature and pressure were lower. In these new conditions, lots of crystals began forming at the same time, reducing the amount of space each one had to grow. These smaller crystals of the matrix bumped into each other as they formed, so they don't have well-developed crystal shapes.

The large, well-developed plagioclase feldspar crystals began forming before the rest of the magma solidified. Zoning is evident in the large crystal in the center of the photo. Each zone reflects different conditions in the magma as the crystal grew.

is granite has large crystals of hornblende, which began
rming long before the surrounding matrix crystallized.

Granite with large crystals of potassium feldspar.

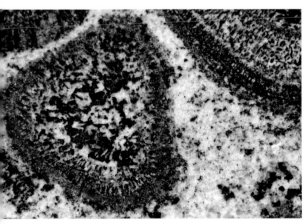

bicules are strange, round balls in igneous rocks
at can look like onions, with alternating layers
light and dark minerals. How they form remains
mystery. What geologists do know is that every
bicule starts with a core—either a crystal, a
nolith, or a chunk of magma that solidifies
rly in the magma's cooling history. Magma
gins solidifying around the core, but conditions
ange suddenly and repeatedly, affecting which
nerals crystallize. The mechanism that causes
e changing conditions isn't understood, and
ere may be many mechanisms.

Orbicules in the Idaho Batholith mystify hikers and geologists alike.
—JAMES LAINSBURY PHOTO

Veins and dikes cross a cliff face in the Black Canyon of the Gunnison in Colorado.

The Broth

By the time most of a granite has crystallized, the remaining magma is very fluid rich and may have high concentrations of less common elements, such as lithium, tantalum, fluorine, boron, beryllium, gold, copper, and zinc. The elements can move around easily in the remaining fluid, which facilitates the growth of large and unusual minerals. This hot fluid can circulate through cracks in the solidifying granite, filling them with veins of minerals. Veins look like dikes, but dikes are bigger and form from intruding magma, not from the fluid that remains after an igneous body has nearly solidified. If you see light-colored lines crisscrossing a cliff face, they're either veins or dikes. Miners often search for gold in veins, looking for the mother lode. Whether a fluid has the elements necessary to create valuable minerals depends on both the composition of the magma and the type of country rock the fluids circulate through.

Veins and dikes made entirely of large crystals are called *pegmatites* and are essentially coarse-grained granite. Pegmatites are usually composed of common minerals, such as quartz, feldspar, and micas, but some pegmatites

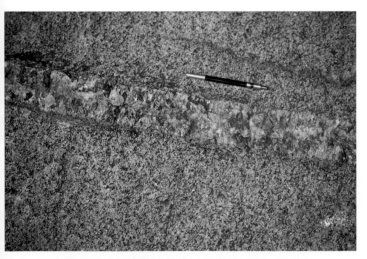

A pegmatite cuts a reddish pink granite in Wisconsin.

A pegmatite cuts granite in this polished specimen. Note the smaller crystals along the edges of the dike where the magma solidified quickly against the cool country rock.

A pegmatite in the foreground with Grand Teton in the background.

also contain crystals of unusual minerals, such as beryl, tourmaline, topaz, and apatite. For a rock to be considered pegmatite, most of its mineral grains must be at least $\frac{1}{2}$ inch across, and in many cases, they're much larger. In a pegmatite mine in the Black Hills in South Dakota, miners found an enormous 40-foot-long crystal of spodumene, a rare mineral that contains lithium.

The impressions left by two spodumene crystals that were removed from a pegmatite mine in the Black Hills of South Dakota. One was 40 feet long. Note the miner at the right center of the photo. —U.S. GEOLOGICAL SURVEY PHOTO, CIRCA 1904

The minerals potassium feldspar and quartz can grow simultaneously in such a way that their crystalline structures are intertwined, like woven threads. This growth form is called graphic granite *because when the rock is cut perpendicular to the long axis of the quartz crystals, the surface is covered with wedges that look like ancient Arabic writing. Graphic granite is usually found in pegmatites.*

Why Is Some Granite Hard and Some Crumbly?

You can hit some granite with a hammer and not make a dent (though you will feel the reverberations throughout your entire body). You can pick other granite apart with your fingers. To understand why this is, we must explore what happens to granite while it's still beneath Earth's surface.

Fractures often form in granite when it's underground. Some form as the magma cools and contracts, others as a result of stresses placed on the granite during mountain building, and still others when the rock above the granite is removed by erosion. When a body of granite is deep below the surface, the confining weight of the rocks above prevent any fractures from opening. There's no wiggle room under that much pressure. Deeply buried granite is also warmer than it would be at the surface, so it's a little more pliable.

Fractures in rock fall into several categories: joints, faults, and cracks. Joints are fractures that often form in response to stress. There are usually multiple joints, all parallel, resembling the slices in a loaf of bread. Faults

Weathering and erosion along parallel fractures—a set of joints—formed these fins of granite.

Two sets of parallel joints, one vertical and one horizontal, control the weathering of this granite tower.

are a different type of fracture in which the rock on one side has moved relative to the other side. And any fracture that isn't a joint or a fault is a crack. Got it?

One type of fracture unique to granitic rock, called an *exfoliation joint*, forms as the rocks above the granite are gradually removed by erosion, which decreases the pressure on the granite. The granite also cools as it gets closer to Earth's surface. Most things change shape when pressure is released or the temperature drops. For instance, if you inflate a basketball indoors where it's warm and then take it outside on a cold day, the ball becomes less firm and loses some of its bounce. As the air inside the ball cools it occupies less space, so the pressure exerted on the inner surface of the ball is reduced. Because granite is solid rock, it can't shrink or expand gradually and evenly, like a basketball. Instead, it responds to a reduction in pressure and temperature by cracking, just as pottery sometimes cracks when the kiln cools.

Exfoliation joints form parallel to the surface of the land because the granite expands upward as overlying rock is removed and pressure is released. A set of curved exfoliation joints resembles the layering of an onion. Exfoliation joints and other fractures provide openings for water to move through granite and begin the weathering process.

In the following discussion, you'll read a lot about weathering and erosion, two processes that work together to shape the landscape we see. *Weathering* refers to changes in the color, texture, or composition of rocks on or near the surface of Earth when they're exposed to the air and water of the atmosphere. A gravestone weathers, as does the wood on the deck of a boat. *Erosion*, a more general term that includes weathering, also refers to the breakup and removal of rock. Wind, flowing water, and ice erode rock, removing particles (or sometimes large chunks!) and depositing them elsewhere. For example, runoff erodes topsoil from fields, and waves erode beaches.

The pattern of the exfoliation joints in these granite domes resembles the layers of an onion. The curved joints formed when pressure was released as the load of rock above the granite eroded away.

Breaking Up Is Hard to Do

Water wouldn't be able to do much damage if it only reached the outer surface of a huge body of granite. But rainwater and snowmelt seep into fractures and begin the weathering process long before the granite ever sees the light of day. When exposed to water, the hornblende, micas, and feldspar minerals in granite change to clay minerals. Plagioclase feldspar and biotite are the most susceptible to this alteration, and only quartz is immune. Because of the chemical changes that occur during the transformation, the clay occupies more space than the original minerals, so it pushes against the surrounding minerals, such as quartz, breaking the granite apart. Have you ever picked up a chunk of granite and had it crumble in your hands? This "rotten" granite falls apart into a loosey-goosey substance called *grus* (rhymes with *goose*), which is the German word for "grit." Grus is made up

DOROTHY NORTON

"Breaking up is hard to do."

of small grains of granite, most of them larger than sand and usually composed of at least several minerals still clinging together.

While the granite is still below the land surface, the weathered material remains in place, covering the unweathered granite. This zone of weathered rock, called a *weathering mantle*, can be a just a few feet thick or as much as 300 feet thick. The thickness depends on how much water is available, how wet and warm the climate is, and how quickly the weathered granite at the surface is eroding as it's carried away by water, wind, and ice.

The weathering mantle isn't 100 percent grus. Weathering takes place along fractures, so at first zones of weathering surround solid blocks of rock. Because corners have three sides, they weather faster than flat faces, which have just one side. So as corners wear away, the blocks become rounded boulders. These remnants of solid granite, called *core-stones*, are

Granite breaks down into smaller and smaller pieces on the way to becoming grus (right).

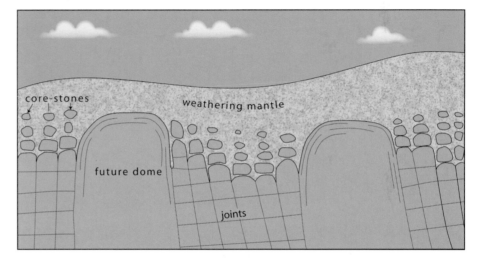

Weathering along sets of parallel joints forms a weathering mantle. Erosion of the weathering mantle exposes domes and core-stones.

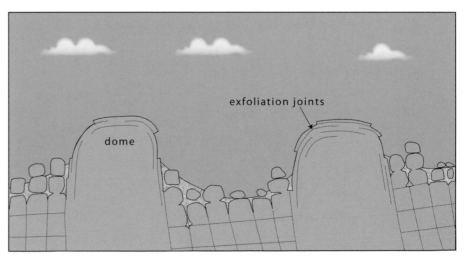

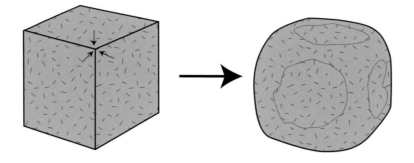

Corners of granite blocks weather faster than flat surfaces because weathering attacks all three sides of a corner.

also part of the weathering mantle. Many large granite boulders that sit on the northern prairies and other formerly glaciated terrains far from the landscape where they originated are granite core-stones. Massive continental ice sheets, working like giant conveyor belts, plucked them from their homes and carried them long distances before dropping them.

Rocks weather at different rates depending on the rock type. For example, shale weathers faster than granite. You can observe this firsthand where veins crisscross granite or granite encloses pieces of country rock. If the vein or country rock weathers more rapidly than the granite, then depressions form as they wear away. If the vein or country rock doesn't weather as fast as the granite, it will stick out of the rock like tiny ridges. Climate also affects the rate of weathering, as rocks in rainy areas weather faster than rocks in dry climates. And coarse-grained granites weather faster than fine-grained granites. For example, the weathering of a single feldspar crystal in coarse-grained granite affects a much larger area than the weathering of a single crystal in fine-grained granite.

If granite weathers in so many ways, you may be wondering how large granite domes and mountains remains intact. In large part, it has to do with the quantity of fractures. Large domes often form in parts of a granite pluton or batholith that don't have as many fractures as other parts. Granite with fewer fractures can better resist the effects of water and ice.

The weathering of a granite body below the land surface can be so thorough that all you'll find when it's finally exposed is a core-stone of granite completely surrounded by grus.

If you break a weathered cobble of granite, you may see evidence of how weathered the rock is. The outer, orangish layer, called a *weathering rind*, is an iron oxide stain left behind by water that has penetrated the rock.

These dark xenoliths stick out because they're more resistant to weathering than the surrounding granite.

This granite pegmatite stands out in relief in places, perhaps because it's slightly more resistant to weathering than the surrounding rock.

At the land surface, rainwater carries the small clay particles of the weathering mantle away, leaving behind grains of quartz. Quartz doesn't break down into another mineral, no matter how much water it's exposed to. Quartz will be quartz forever if it stays on Earth's surface, though the individual grains wear down over time. You'll find quartz sand at the base of most granite outcrops. The sharp edges of the quartz grains wear down as wind blows them this way and that and as water tumbles the grains against rocks and each other. By the time quartz sand has been deposited onto beaches and dunes, where it ultimately ends up, the grains are round and soft enough to walk on.

Much of the hard granite we see in North America has been "cleaned" of its weathered surface. During the ice ages of the past 1.8 million years, glacial ice scraped the surface of much of the granite in Canada, the northern United States, and high mountains farther south, such as the Sierra Nevada. The ice carried away the grus and loose grains of the weathering mantle, uncovering the hard granite beneath.

Rivers also have a knack for scouring away loose stuff and leaving behind smooth, hard rock. And when water freezes and expands in cracks in rocks, sections of rock can break free, exposing fresh unweathered surfaces.

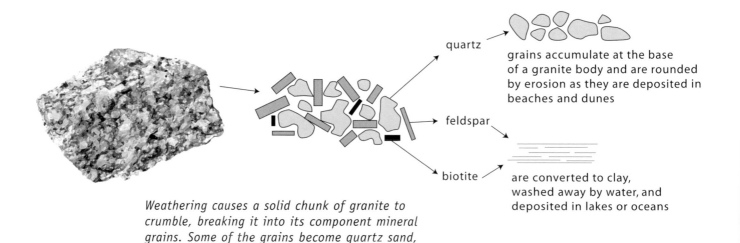

quartz

grains accumulate at the base of a granite body and are rounded by erosion as they are deposited in beaches and dunes

feldspar

biotite

are converted to clay, washed away by water, and deposited in lakes or oceans

Weathering causes a solid chunk of granite to crumble, breaking it into its component mineral grains. Some of the grains become quartz sand, and the rest eventually turn into clay.

This can occur on a very large scale, with large blocks breaking free from a mountainside. Hard rock is also exposed in areas where erosion at the surface is progressing faster than underground weathering. Wind and water remove the weathered material from the rock surfaces and fractures faster than new loose material is produced, leaving behind fins, rounded domes, and large boulders. In general, hard granite hasn't weathered as much as crumbly granite, but newly exposed, hard granite will eventually weather and become crumbly.

Rocks embedded in the moving ice of a glacier scratched this granite in the Sierra Nevada, leaving distinct lines that parallel the direction the ice moved.

Granite at Joshua Tree National Park, now exposed as rounded domes and boulders, probably weathered underground.

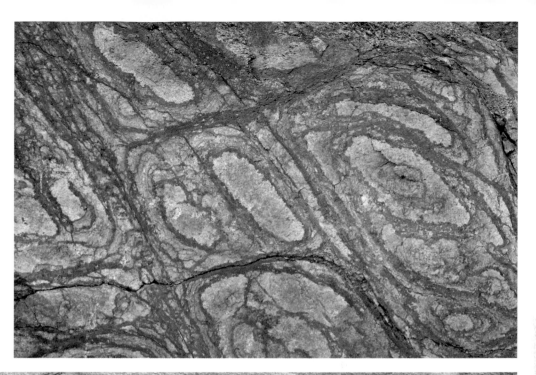

When water flows along fractures where granite has begun to weather, it leaves behind concentric rings of reddish brown iron oxide staining. Called liesegang rings (pronounced leez-ah-gang), they may mark different stages of the infiltration of water into the porous surface of a weathered rock. The stain is most highly concentrated along fractures.

You'll find coarse sand at the base of most granite outcrops.

WEATHERING IS THE PITS

Not many plants can grow on granite. Most rain that falls on it runs off quickly, leaving behind a dry, inhospitable surface. However, some plants do find footholds in cracks or places where seeps provide a steady supply of water. Other plants inhabit shallow depressions called *weathering pits*. The pits deepen as water accumulates and accelerates the weathering process. Eventually soil collects in the depressions, providing sufficient nutrients for small plants. The soil forms from particles washed or blown into the pit, including clay eroded from the granite and organic matter. Even lichens growing on granite can trap dust and begin the slow process of soil development.

Weathering pits in granite Stone Mountain, Georgi

Depressions in granite can weather into pits. Runoff carries small particles of clay, sand, and organic material into the pit, making it hospitable to plants.

A small pine grows from a crack in granite at Stone Mountain, Georgia. Note the dark stains left by water flowing over the rock.

Plants grow in cracks in granite in the Wasatch Range, Utah. Water is channeled into the cracks, which intensifies weathering and leads to soil development.

How Old Is Granite?

Like all rocks that crystallize from magma, granite can be dated, no matter how old it might be. The oldest known rock in the world is a 4.28-billion-year-old metamorphosed volcanic rock in northern Quebec. Grains of the mineral zircon found in sandstone in Australia have been dated at 4.2 billion years. The rock they eroded from before being integrated into the sandstone hasn't been found and may no longer exist. Zircon is a minor mineral in igneous rocks and is very stable. Once it forms it sticks around for a long time and doesn't change unless it's metamorphosed. There may be even older rocks or mineral grains, but geologists have yet to find them.

Scientists estimate Earth is about 4.6 billion years old—4,600,000,000 years. That number can be hard to fathom. If you compare the age of Earth with a single year, with Earth forming on New Year's Day, then the dinosaurs went extinct on December 25 and our species, *Homo sapiens*, came to the New Year's Eve party around 11:00 p.m. Recorded human history fits within the last minute of the year, but geologic history spans the entire calendar year. Being able to tell the age of igneous rocks, which were forming in one place or another throughout much of Earth's history, has helped geologists piece together the calendar of geologic time.

For example, clusters of granite bodies of about the same age in the same region tell geologists that some tectonic event was causing granite to form at that time and place in the past. Numerous bodies of granite in the western United States, from the Sierra Nevada to the Idaho Batholith, formed between 200 and 70 million years ago. It's likely that the western margin of North America was colliding with one or more tectonic plates to its west during this interval of time.

Just as people have telltale signs of age, such as wrinkles and gray hair, so do igneous rocks, but you can't see them. To date an igneous rock, geologists look at radioactive isotopes, which are different varieties of a given element. It's sort of like genus (a grouping of animals with similar characteristics) and species (a type of animal within a genus) in the world of biology. Within the genus *Canis*, there are several types of doglike animals. If *Canis*

Linear bands of granite in the western United States formed deep below the surface between 200 and 70 million years ago when the western edge of the continent was a subduction zone.

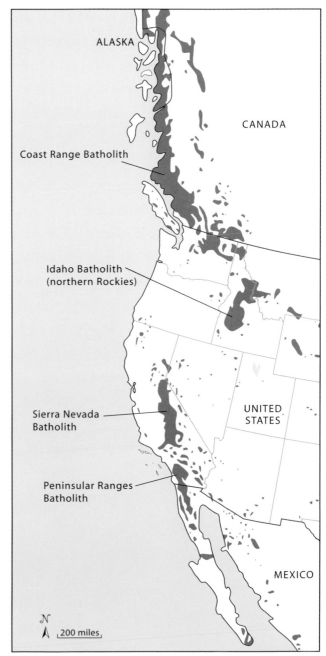

were an element, the different species (coyotes, jackals, wolves, and dogs) would be the isotopes. But unlike dogs and their kin, some isotopes are radioactive, which means they're unstable and change over time. Through a process known as *radioactive decay*, atoms of a radioactive isotope of one element change into atoms of a stable isotope of a different element (called its *decay product*), usually by shedding a piece of themselves. Scientists have determined that radioactive isotopes decay at specific rates. And because decay starts as soon as the mineral crystallizes, determining the ratio of original isotopes to new isotopes will reveal how long ago the mineral formed.

Several minerals in granite contain radioactive isotopes. One of the most useful minerals for dating old rocks is zircon, which contains an isotope of uranium that decays to an isotope of lead. It takes 4.5 billion years for half of the uranium atoms to decay to lead atoms. This time interval is called the *half-life*. For example, if there were one hundred radioactive uranium atoms in a zircon mineral when it formed, after 4.5 billion years, there would be fifty remaining uranium atoms and fifty new lead atoms. Half of the remaining radioactive isotopes would decay in another 4.5 billion years, and so on. A rock doesn't have to be 4.5 billion years old to be dated. Scientists just determine the ratio of the radioactive isotope to its decay product.

Dark zircon crystals in granite contain radioactive isotopes that decay at known rates, allowing geologists to determine the age of the granite. These zircon crystals are unusually large; most are much smaller.

Dating in the Dark Ages

The ability to get a numerical age for crystalline rocks revolutionized geology in the early twentieth century. For the first time, geologists could say, "This rock is 50 million years old," rather than just, "This rock is older than that rock." Prior to isotopic dating, geologists could only determine relative ages, such as that sedimentary rocks were younger than the rocks below them, and that bodies of granite were younger than the rock they intruded. They could also determine the relative ages of some rocks by studying the fossilized remains of animals and plants the rocks contained. For example, they could deduce that a sedimentary rock with a very characteristic fossil was probably the same age as another rock containing the same fossil, even if the two rock types were separated by a great distance. But this didn't allow them to determine the age of the rock in years.

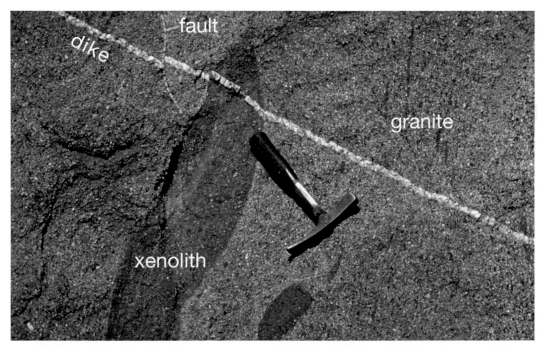

What happened first in this rock outcrop? The dark xenolith is the oldest rock. It was plucked from country rock by granitic magma, which solidified into the second oldest rock. Next, the small white dike intruded both the granite and the xenolith and was later broken by a fault. Using these types of observations, geologists can assign relative ages to rocks.

Fossils and their location in layers of rock, and whether a rock is above or below another, remain important in determining the relative ages of sedimentary rocks. But if granite is present within or below the sedimentary rock, geologists can determine a much more accurate age range using isotopic dating. The numeric age allows geologists to give the sedimentary rock a minimum or maximum age. If an 80-million-year-old body of granite intruded the sedimentary rock, they know the sedimentary rock must be more than 80 million years old. And if sediment was deposited on top of the granite, they know that the sediment is less than 80 million years old.

The younger sandstone (above) *was deposited on top of a deeply weathered, crumbly granite* (below). *The sandstone was deposited in horizontal layers on the flat surface of the granite, and later the whole sequence of rocks was tilted during some tectonic event.*

The younger, light-colored dike of granite intruded older, darker gneiss.

HALFTIME

Half-life is the time it takes for half of the atoms of a radioactive isotope to decay and become atoms of an isotope of another element. Imagine that a penny is an atom, with the head being the original radioactive atom, and the tail the new, stable atom. Put sixty-four pennies in a cup and shake them up. Pour the pennies onto a table and remove the tails. Put the pennies that are heads up back in the cup, shake it again, then pour them out and remove the tails once again. If exactly half of the pennies are tails up each time, and you repeat this until all of the pennies are gone, you'll need to shake the pennies seven different times—seven half-lives—to remove all of them. However, there is a certain amount of randomness to this exercise, so it's unlikely you'll get exactly half heads and half tails each time you shake the pennies. But if you do this exercise multiple times and average your results, you'll find that it typically takes about seven times to remove all the pennies from the cup. If the shaking represents a half-life of 50 years, it would take 350 years (7 times 50) for all the radioactive atoms (the penny heads) to decay to stable atoms (the penny tails).

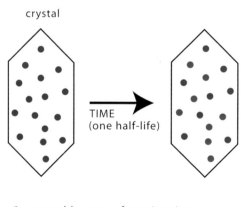

crystal

TIME
(one half-life)

● unstable atom of uranium isotope

● stable atom of lead isotope

Over time, atoms of an unstable isotope in a mineral decay, becoming atoms of a stable isotope. Analyzing the ratio between the two allows scientists to determine the age of rocks.

Appendix
Where Can You Find Granite in the United States?

This list includes national, state, county, and city parks and public lands. It is not comprehensive but includes some of the better sites to see granite. In states with a lot of granite, such as California, we mainly listed major parks. In states with fewer outcrops, we listed some out-of-the-way sites. Keep in mind that although granite outcrops occur at all these sites, not all rocks at a listed site are granite. We listed towns with *granite* in their name but did not list every topographic feature in the United States with *granite* in its name.

ALABAMA
Flat Rock Park in Wedowee

ALASKA
Chena River State Recreation Area: granite domes
Denali National Park and Preserve: Alaska Range has a granite core
Glacier Bay National Park and Preserve: metamorphic sedimentary rocks intruded by granite
Katmai National Park and Preserve: granite plutons
Klondike Gold Rush National Historical Park
Lake Clark National Park and Preserve: granite plutons
Misty Fiords National Monument: Coast Range has a granite core

ARIZONA
Agua Fria National Monument
Catalina State Park: Mount Lemmon is granite
Grand Canyon National Park: Zoroaster granite
    intrudes older schists and gneisses
Granite Basin Recreation Area and Granite Mountain Wilderness
Granite Dells near Payson
Granite Dells north of Prescott
Hualapai Mountain Park
Kitt Peak National Observatory
Oracle State Park

ARKANSAS

Place Names and Other Notable Facts

Granite Mountain in Gillam Park in Little Rock is syenite, an intrusive
igneous rock composed mainly of feldspar but not enough quartz to
be called *granite*

CALIFORNIA

Alabama Hills Recreation Area
Anza-Borrego Desert State Park: Peninsular Ranges Batholith
Castle Crags State Park
Cuyamaca Rancho State Park: Peninsular Ranges Batholith
Death Valley National Park
D. L. Bliss State Park: Sierra Nevada Batholith
Garrapata State Park: Hobnail granite
Joshua Tree National Park
Mojave National Preserve: Granite Mountains
Mount San Jacinto State Park: Peninsular Ranges Batholith
Point Reyes National Seashore
Saddleback Butte State Park
Sequoia and Kings Canyon National Parks: Sierra Nevada Batholith
Trinity Alps Wilderness
Whiskeytown National Recreation Area: Shasta Bally Batholith
Yosemite National Park: Sierra Nevada Batholith

Place Names and Other Notable Facts

"Black granite" of San Diego County is gabbro
Granite Bay (town and bay)
Graniteville (town)

COLORADO

Black Canyon of the Gunnison: mostly gneiss but some granite
Golden Gate Canyon State Park
Mount Evans Wilderness Area: Mount Evans Batholith
Mueller State Park: Pikes Peak Batholith
Pike National Forest: Pikes Peak Batholith at Devils Head,
along South Platte River, at Pikes Peak
Rocky Mountain National Park: Longs Peak–St. Vrain Batholith
Royal Gorge on Arkansas River: Pikes Peak Batholith

COLORADO continued

Place Names and Other Notable Facts
Granite (town)
Most Front Range mountains in Colorado contain granite

CONNECTICUT

Lots of granite gneiss but very little unmetamorphosed granite

DELAWARE

Place Names and Other Notable Facts
"Blue granite" quarried in Delaware is gneiss

FLORIDA

GEORGIA

Heggie's Rock National Natural Landmark in Columbia County
Panola Mountain Conservation Park
Stone Mountain Park

Place Names and Other Notable Facts
Elberton, which has the Elberton Granite Museum, bills itself
    as the Granite Capital of the World (see also Barre, Vermont)

HAWAII

IDAHO

Castle Rocks State Park
City of Rocks National Reserve
Frank Church–River of No Return Wilderness Area: Idaho Batholith
Sawtooth Wilderness: Idaho Batholith
Selkirk Mountains: Kaniksu Batholith
Selway-Bitterroot Wilderness Area: Idaho Batholith

Place Names and Other Notable Facts
Granite (town)

ILLINOIS

Place Names and Other Notable Facts

Granite City (town named for enameled steel called Granite Ware
that was manufactured there)

INDIANA

IOWA

KANSAS

KENTUCKY

LOUISIANA

MAINE

Acadia National Park: Cadillac Mountain granite
Baxter State Park: Katahdin granite
Grafton Notch State Park: Bear River cuts a gorge through granite
Moosehorn National Wildlife Refuge: Meddybemps granite
mixed with basalt
Mount Waldo quarry
Saint Croix Island International Historic Site: Red Beach granite
Sebago Lake State Park: Sebago granite
Vinalhaven Island: Vinalhaven granite exposed at town parks

MARYLAND

Chesapeake & Ohio Canal National Historical Park:
Dalecarlia granite near milepost 6.5
Gunpowder Falls State Park: Gunpowder granite
Patapsco Valley State Park: Ellicott City granodiorite
exposed in part of park near Ellicott City

Place Names and Other Notable Facts

Granite (town)

MASSACHUSETTS
Halibut Point State Park: Cape Ann granite
Fort Phoenix State Reservation
Quincy Quarries Reservation: Quincy granite

Place Names and Other Notable Facts
Graniteville (town)

MICHIGAN
Marquette Tourist Park: granite intrudes greenstone,
    a metamorphic rock, along the Dead River

MINNESOTA
Big Stone National Wildlife Refuge: Ortonville granite
Boundary Waters Canoe Area Wilderness: Saganaga Batholith
Stearns County Quarry Park and Nature Preserve
Voyageurs National Park: Vermilion Granitic Complex
    of gneiss, migmatite, and granite

Place Names and Other Notable Facts
Granite Falls (town named for 3.5-billion-year-old granite gneiss)

MISSISSIPPI
none

MISSOURI
Elephant Rocks State Park: Graniteville granite
Hawn State Park

Place Names and Other Notable Facts
Graniteville (town)
"Missouri Red" is the commercial name for Graniteville granite

MONTANA
Absaroka-Beartooth Wilderness: volcanic rocks, gneiss, and granite
Beaverhead National Forest: Pioneer Batholith of the Pioneer Range
Deerlodge National Forest: Boulder Batholith at Homestake Pass
Humbug Spires Wilderness Study Area: Boulder Batholith
Lolo National Forest: Lolo granite near Lolo Hot Springs

Place Names and Other Notable Facts
Granite Peak, the highest point in Montana, is gneiss

NEBRASKA

NEVADA

Lake Tahoe State Park: Sierra Nevada Batholith
Prison Hill Recreation Area in Carson City: granite and volcanic rock
Ruby Mountain Scenic Area: gneiss intruded by granite

Place Names and Other Notable Facts

Highest point in Nevada is granite of Boundary Peak
Many of Nevada's mountain ranges are peppered with granite intrusions,
   including Granite Peak in the Santa Rosa Range

NEW HAMPSHIRE

Cardigan State Park: primarily granodiorite
Crawford Notch State Park: Willard Peak is Conway granite
   of White Mountain Batholith
Franconia Notch State Park: White Mountain Batholith

Place Names and Other Notable Facts

Granite (town)
Granite is the New Hampshire state rock
New Hampshire is the Granite State

NEW JERSEY

Round Valley State Recreation Area: granite on
   western shore of reservoir
Schooleys Mountain County Park: Lake Hopatcong granite
Wanaque Wildlife Management Area: Byram granite
Wawayanda State Park: granite pegmatite on north side of park road

Place Names and Other Notable Facts

Pompton pink granite was used for the landing at the entrance
   of the Smithsonian National Museum of Natural History

NEW MEXICO

Capitan Mountains Wilderness Area
Organ Mountains (east side) accessed via Aguirre Springs Campground
Sierra Blanca Range
Tres Piedras crags (west of Taos)

NEW MEXICO continued

Place Names and Other Notable Facts
Many mountains have granite cores, including the Sandia Mountains,
 Zuni Mountains, Taos Mountains, and Tusas Mountains

NEW YORK

Place Names and Other Notable Facts
Graniteville (town)
Peekskill granite is quarried near Peekskill

NORTH CAROLINA
Boones Cave Park: granite of Churchland Plutonic Suite
Medoc Mountain State Park: Butterwood Creek granite
Mitchell Mill Natural Area in Wake County: Rolesville
 Batholith (also exposed at Temple Flat Rock Preserve,
 Clemmons Educational State Forest, and near the towns of
 Lizard Lick and Rolesville)
Pisgah National Forest: Looking Glass pluton exposed at
 Looking Glass Rock
South Mountains State Park: Walker Top granite
Stone Mountain State Park

Place Names and Other Notable Facts
Granite Falls (town)
Granite Quarry (town named for quarries in Salisbury granite)
Granite is North Carolina state rock

NORTH DAKOTA
none

OHIO
none

OKLAHOMA
Arbuckle Mountains: Tishomingo granite exposed,
 particularly at "10 Acre Rock" near Troy
Great Plains State Park
Quartz Mountain State Park: Quanah granite and Mt. Scott granite
 of the Wichita Mountains
Wichita Mountains Wildlife Refuge

OKLAHOMA continued

Place Names and Other Notable Facts
Granite (town)

OREGON

Eagle Cap Wilderness: Wallowa Batholith
North Fork John Day Wilderness: Bald Mountain Batholith
Rogue River National Forest: Mount Ashland pluton

Place Names and Other Notable Facts
Granite (town)

PENNSYLVANIA

RHODE ISLAND

Black Point fishing area at Point Judith Neck:
    Narragansett Pier granite
Cliff Walk: Newport granite
Fort Wetherill State Park: Newport granite
Lincoln Woods State Park: Dedham granite
Snake Den State Park: Scituate granite

SOUTH CAROLINA

Forty Acre Rock: Pageland granite in the Forty Acre Rock
    Heritage Preserve near Taxahaw in Lancaster County

Place Names and Other Notable Facts
Graniteville (town)
Winnsboro blue granite is the South Carolina state rock

SOUTH DAKOTA

Bear Butte State Park
Custer State Park
Mount Rushmore National Memorial: Harney Peak granite

Place Names and Other Notable Facts
Red granite is quarried in Milbank and Big Stone City
    near the Minnesota border

TENNESSEE

Roan Mountain State Park: Beech granite

TEXAS

Enchanted Rock State Natural Area: Town Mountain granite
Granite Mountain: quarry for Texas pink granite
Hueco Tanks State Park and Historic Site: syenite,
    an intrusive igneous rock composed mainly
    of feldspar but not enough quartz to be called *granite*

Place Names and Other Notable Facts

Granite Shoals (town)

UTAH

Devils Playground in Grouse Creek Mountains: Emigrant Pass pluton
G. K. Gilbert Geologic View Park: A mountain of quartz monzonite,
    a cousin of granite, is visible from the park, which is located
    near the mouth of Little Cottonwood Canyon southeast of
    Salt Lake City
Granite Creek cuts through granite in Deep Creek Range
Pine Valley Mountains Wilderness
Rock Corral Recreation Area in Mineral Mountains
    (includes Granite Peak)

Place Names and Other Notable Facts

Granite (town)

VERMONT

Groton State Forest: Knox Mountain pluton in
    New Hampshire Plutonic Series
Rock of Ages Quarry: Barre granite of New Hampshire Plutonic Series
Willoughby State Forest: Willoughby pluton in
    New Hampshire Plutonic Series

Place Names and Other Notable Facts

Barre bills itself as the Granite Capital of the World
    (see also Elberton, Georgia)
Granite, along with marble and slate, are the Vermont state rocks
Graniteville (town)

VIRGINIA

Belle Isle State Park: Petersburg granite
Shenandoah National Park: Old Rag Mountain is granite

WASHINGTON

Alpine Lakes Wilderness: Mount Stuart Batholith
Little Pend Oreille National Wildlife Refuge:
    Crystal Falls spills over granite
Mount Rainier National Park: some granite in a mostly volcanic region
Mount Spokane State Park
North Cascades National Park
Selkirk Mountains: Kaniksu Batholith

Place Names and Other Notable Facts

Granite Falls (town)

WEST VIRGINIA

WISCONSIN

Baxter's Hollow State Natural Area: Baxter Hollow granite
    exposed below quartzite along Otter Creek
Chute Pond County Park south of Mountain: Wolf River Batholith
Montello Granite Quarry Waterfalls Park: Montello granite
Tigerton Dells on Embarrass River: Wolf River Batholith

Place Names and Other Notable Facts

Redgranite (town)
Red granite is Wisconsin's state rock
Rib Mountain of the Granite Peak Ski Area is quartzite,
    a metamorphic rock, and not granite

WYOMING

Cloud Peak Wilderness Area in the Bighorn Mountains
Grand Teton National Park: granite and gneiss
Granite Mountains (Sweetwater Hills/Split Rock area)
Independence Rock State Historic Site: granite forms this
    Oregon Trail landmark; Sweetwater River cuts through
    granite at nearby Devils Gate
Popo Agie Wilderness Area in the Wind River Range
Vedauwoo Recreation Area: Sherman granite of the Laramie Mountains

Glossary

basalt. A black or dark gray extrusive igneous rock that consists mainly of tiny crystals of plagioclase feldspar and pyroxene. Basalt has the same composition as gabbro but much smaller mineral grains because it cooled from lava on the surface of Earth.

batholith. A large body of igneous rock, up to 100 miles long and 50 miles wide, made up of many plutons.

bedrock. The hard rock at Earth's surface that underlies soil and unconsolidated sediments.

biotite. A brown to black mineral in the mica group of minerals that breaks along one surface, forming flakes.

core-stone. A boulder of hard rock entirely enclosed by weathered rock.

country rock. The rock that magma intrudes.

crust. The outer layer of Earth, composed of hard, brittle rock.

crystal. The shape, bound by flat surfaces called *faces*, that forms as a mineral grows. Each mineral has a characteristic crystal shape.

dike. A narrow intrusion of igneous rock that cuts through country rock.

diorite. One of the three main intrusive igneous rocks. It contains less silica and more dark minerals than granite.

dome. A large, rounded mass of rock.

element. In chemistry, a substance that consists of atoms of only one kind.

erosion. The wearing away or slow destruction of the landscape by water, wind, or ice.

fault. A fracture in which rock on one side has moved relative to rock on the other side.

feldspar. A mineral group that includes potassium feldspar and plagioclase feldspar, two main minerals of granite. These light-colored minerals break along two surfaces and look boxy.

fin. A narrow landform that resembles the fin of a fish.

gabbro. One of the three main intrusive igneous rocks. It contains less silica and more dark minerals than granite and diorite. It is the chemical equivalent of basalt, an extrusive igneous rock.

gneiss. A metamorphic rock with bands, or layers, of light and dark minerals.

grus. Grains of disintegrated granite ranging in size from sand to gravel.

hornblende. An iron- and magnesium-bearing, brown to black mineral of the amphibole group of minerals. It is one of the dark minerals in granite.

igneous. One of the three main types of rocks. It includes rocks that crystallize from cooling magma. If magma cools deep beneath the surface, it is an *intrusive* igneous rock; if it cools on the surface, it is an *extrusive* igneous rock.

intrusion. A body of igneous rock formed from magma that cooled below Earth's surface.

isotope. A different species of atom of the same chemical element.

joints. Planar, parallel fractures in hard rock. Joints of the same orientation are called a *set*. Joints in granite that look like the layers of an onion are called *exfoliation joints*.

lava. Magma on the surface of Earth.

magma. Melted rock below the surface of Earth.

mantle. The rock that lies below Earth's crust. Rock in the mantle can flow like taffy. The mantle is the source of some magma and is also responsible for the heat that melts the crust into magma.

matrix. The material in which something is enclosed.

metamorphic. One of the three main types of rocks. Heat and pressure change igneous or sedimentary rocks to metamorphic rocks.

mica. A group of minerals that break along one surface and form flakes. They are minor minerals in granite.

migmatite. A rock that is a mixture of igneous rock and country rock.

mineral. A naturally occurring inorganic substance that has an orderly arrangement of atoms. Rocks are made of minerals.

molecule. The smallest particle of a substance composed of more than one atom.

muscovite. A mineral in the mica group of minerals.

outcrop. An exposure of bedrock.

pegmatite. A dike or vein with large mineral crystals.

plate tectonics. The theory that Earth's crust consists of plates that move relative to each other.

pluton. An irregularly shaped body of intrusive igneous rock.

precipitate. The process by which a substance comes out of solution and becomes a solid.

quartz. A mineral that is 100 percent silica. It is one of the major light-colored minerals in granite.

radioactive decay. A process in which atoms of a radioactive, or unstable, isotope of an element change into atoms of a stable isotope of a different element over time.

sandstone. A sedimentary rock consisting of sand-size grains of sediment held together by a natural cement.

sediment. Loose grains of rock that are not cemented together.

sedimentary. One of the three main types of rocks. It includes rocks formed at Earth's surface by the deposition of sediments by water or wind or the precipitation of minerals from water.

silica. A molecule composed of one atom of silicon and two atoms of oxygen.

silicate. A mineral composed of a combination of silica and other elements.

sill. An igneous intrusion that forms between layers in country rock.

spreading zone. A tectonic margin where tectonic plates spread apart and magma rises to the surface.

subduction zone. A tectonic margin where one tectonic plate descends beneath another.

tectonic margin. Where tectonic plates move away from each other or collide, or one plate descends into the Earth under another plate.

tectonic plates. The individual sections of Earth's crust that move independently of each other.

texture. The surface appearance of a rock.

vein. A mineral-filled fracture.

volcano. An opening at Earth's surface from which heat, gases, and magma escape from the interior.

weathering. The processes by which water and air change the color, texture, and composition of rocks on or near the surface of Earth.

xenolith. A piece of country rock enclosed within an igneous rock.

References

Chen, Guo-Neng, and Rodney Grapes. 2007. *Granite Genesis: In-situ Melting and Crustal Evolution*. Dordrecht, the Netherlands: Springer.

Levin, Harold L. 2003. *The Earth Through Time*. Seventh Edition. San Francisco: John Wiley & Sons.

Marshak, Stephen. 2005. *Earth: Portrait of a Planet*. Second Edition. New York: W. W. Norton & Company.

Migon, Piotr. 2006. *Granite Landscapes of the World*. University of Oxford Press.

Murdy, William H., and M. Eloise Brown Carter. 1999. *Guide to the Plants of Granite Outcrops*. Athens: University of Georgia Press.

Wessels, Tom. 2001. *The Granite Landscape: A Natural History of America's Mountain Domes, from Acadia to Yosemite*. Woodstock, Vermont: The Countryman Press.

*   To compile the appendix, I relied on Web sites of individual state geologic surveys as well as those of the National Park Service, U.S. Geological Survey, and individual state parks.

** For information about the geology of specific regions, see books in the Roadside Geology series and Geology Underfoot series, published by Mountain Press Publishing Company.

Index

Page numbers in bold italics refer to photographs.

About the Authors

Jennifer H. Carey grew up on granitic rock of the Mount Ashland pluton in southern Oregon. Her interest in geology began early while panning for gold in clear streams sparkling with mica, searching for agates on Oregon beaches, and hiking in the mountains. She obtained a BA in geology at Carleton College and an MS in environmental studies at the University of Montana. She has edited geology books at Mountain Press Publishing Company in Missoula, Montana, since 1995.

Marli Bryant Miller started photographing rocks while majoring in geology at Colorado College in the early 1980s. Her passion for geological photography grew as she spent more and more time in the field while attending graduate school at the University of Washington. She completed a PhD in geology in 1992 and started teaching at the University of Oregon in 1997. She regularly contributes geologic images to textbooks and other publications and is the author of *Geology of Death Valley National Park*.